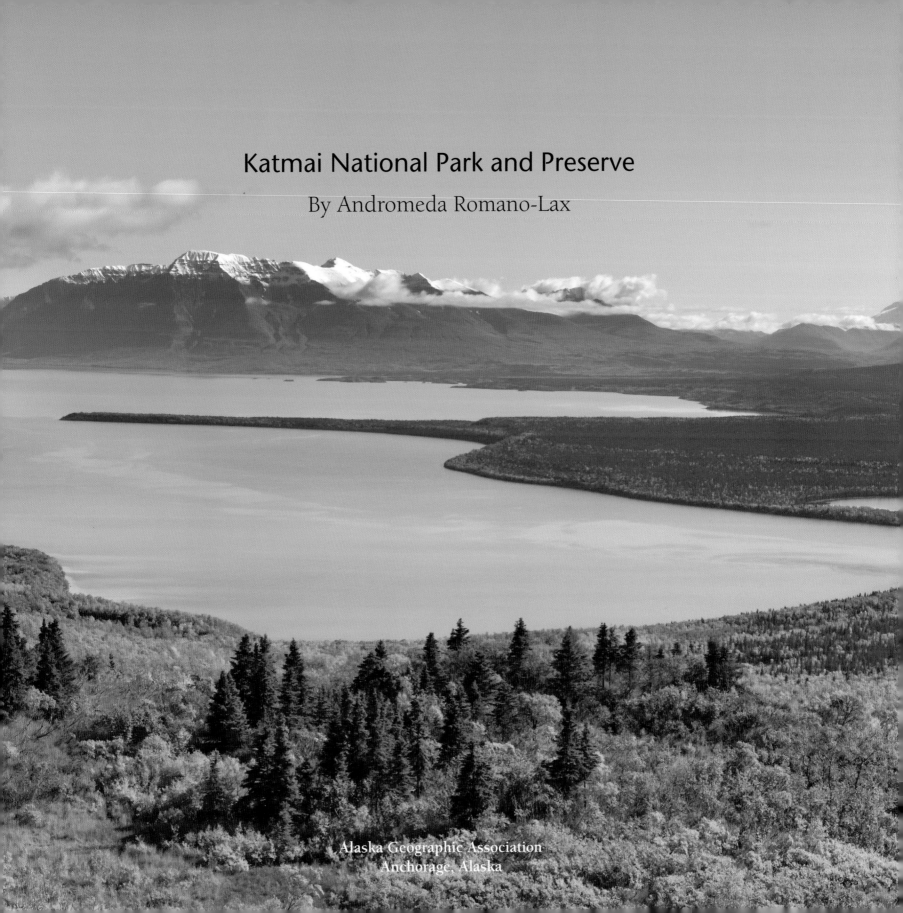

Katmai National Park and Preserve

By Andromeda Romano-Lax

Alaska Geographic Association
Anchorage, Alaska

Alaska Geographic Association thanks Katmai National Park and Preserve for their assistance in developing and reviewing this publication. Alaska Geographic works in partnership with the National Park Service to further public education and appreciation for national parks in Alaska. The publication of books, among other activities, supports and complements the National Park Service mission.

Photography: Alaska Volcano Observatory, US Geological Survey, by Game McGimsey, 17, 18; © Phillip Colla / Oceanlight.com, 37, 56, 58; © Ken Conger, cover; © Jon Cornforth / cornforthimages.com, 28; © Scott Dickerson, 40; © Mark Emery, 29, 36, 49, 53; © Patrick Endres / AlaskaPhotoGrahpics.com, ii-iii, 22, 32-33, 42-43, 45, 48, 59, 63, 67 bottom inset; © Howie Garber / DanitaDelimont.com, 38; © Ron Horn, 61; © Steve Kazlowski / DanitaDelimont.com, 62, 65; © Frans Lanting / www.lanting.com, 20-21; Library of Congress HABS AK, 4-KISAL.V,1-2, 51; QT Luong / terragalleria.com, vi-1, 10, 12, 13; Laura Lohrman Moore / Shutterstock.com, 9; National Park Service, Museum Management Program and Katmai National Park and Preserve, Incised Stone, KATM 41753, 27; National Park Service photo by Jeanne Schaaf, 23; National Park Service photos by Roy Wood 16, 39; Oksana Perkins / Shutterstock.com, 25; © Hugh Rose / AccentAlaska.com, 41; © Hugh Rose / DanitaDelimont.com, 14; Santa Clara University, Department of Archives and Special Collections, H-3256, 19; © Paul Souders / DanitaDelimont.com, 34, 50, 66-67; © David Taylor / SixtyoneNorth.com, iv, v, 54-55, 67 middle inset; © John Tobin / tobinphoto.com, 44; University of Alaska Anchorage, Archives and Special Collections, Consortium Library, National Geographic Society, Katmai Expeditions, 1913-1919, UAA-HMC-0186: E.C. Kolb, 8, 11, 26, D.B. Church, 4-5, 15, J.D. Sayre, 47; US Geological Service photos by G.C. Martin, 2-3, 6; © Kennan Ward, 57; © Art Wolfe / artwolfe.com, 67 top inset

Illustrations: Kristin Link
Map Illustration: Denise Ekstrand
Park Map: Courtesy of the National Park Service
Graphic Design: Debbie Whitecar
Editor: Jill Brubaker
Project Manager: Lisa Oakley
Agency Coordinator: Roy Wood, Katmai National Park and Preserve
ISBN: 978-0-9825765-5-7

© 2012 Alaska Geographic Association. All rights reserved.
Printed in China

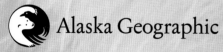

www.alaskageographic.org

Alaska Geographic is the official nonprofit publisher, educator, and supporter of Alaska's parks, forests, and refuges. Connecting people to Alaska's magnificent wildlands is at the core of our mission. A portion of every book sale directly supports educational and interpretive programs at Alaska's public lands. Learn more and become a supporting member at: www.alaskageographic.org

Library of Congress Cataloging-in-Publication Data

Romano-Lax, Andromeda, 1970-
  Katmai National Park and Preserve / By Andromeda Romano-Lax. -- 1st ed.
     p. cm.
  ISBN 978-0-9825765-5-7
  1. Katmai National Park and Preserve (Alaska) 2. Geology--Alaska--Katmai National Park and Preserve. I. Title.
  QE84.K32R66 2012
  557.98'4--dc23
                  2011050714

Cataclysm and Renewal

KATMAI NATIONAL PARK and PRESERVE

Contents

Upheaval and Attraction.. 1

Explosions and Explorers...................................... 4

A Land Both Harsh and Plentiful 20

Surges and Seasons 32

An Industrious and Alluring Frontier..................... 42

The Bears of Katmai.. 54

Upheaval and Attraction

The morning of June 6, 1912 dawned calm, warm, and clear on the remote Katmai coast of southwest Alaska, a suspenseful pause following days of rumbling earthquakes, forecasting the geological drama to come. Then a substance that looked like snow began to fall, blanketing rivers and whitening the mirror-like surface of bays. On the shore it coated the tips of green grass stalks and the heads of wildflowers. Soon, the beachfront was thrumming with a sound of tapping, as if grains of rice were falling—not soft snow at all, but ash, mixed with bigger pieces of pumice. Now the strange sky grew dark as the ash cloud advanced. Thunder and lightning—both rare in this part of the country—added to the strange gloom, until the choking, rock-filled cloud had turned day to a deep, noxious night that would last over 60 hours, created by the largest volcanic eruption of the twentieth century.

Dr. Robert F. Griggs, a National Geographic Society explorer, would later write from the perspective of a non-Alaskan living far from the eruption: "No one suspected that such a catastrophe was in progress until the 6th of June, when volcanic ash suddenly began falling over all of northwestern America." For hundreds of miles, the towering ash cloud could be seen. As the volcanic aftereffects spread—acid rain in Vancouver, a dusting of ash in Seattle, a lowering of temperatures, haze, and brilliant sunsets worldwide—so, too, did curiosity about this remote corner of Alaska. A cataclysm had shaken and transformed the Katmai

coast. But just as surely, this cataclysm had lifted a veil. In years to come, a place that had been obscure to many would become a place of fascination and enduring attraction.

It is a common motif in the inspiring story of America's national parks that a place conserved for one dramatic feature becomes known and cherished later for many reasons—reasons that gain in value as the world changes and as wild places and intact ecosystems become yet more rare. So, too, in Katmai National Park.

Four years after the Novarupta eruption, Griggs and his team of National Geographic Society explorers shared with readers a tantalizing portrait of this geologically violent landscape, a place of smoking fumaroles (now extinct) that Griggs called "The Valley of Ten Thousand Smokes." Several decades later, this place would become known for its world-class fly-in fishing and its congregating brown bears, both eager to plunge into rivers endowed with exceptional runs of sockeye salmon.

The visitor focus may broaden or change, but the foundation remains the same. Nature created this dynamic and fertile landscape: its ash-buried valley and stunningly clear lakes, its fumaroles and its extraordinary fishing grounds. Katmai is now a 4.1-million acre national park and preserve, encompassing not only diverse ecosystems but also cultural landscapes at the heart of a peninsula, where people have been both nourished by nature's bounty and challenged by nature's cataclysms for thousands of years.

Katmai Village, August 13, 1912

Explosions and Explorers

Volcanic ash from the Novarupta eruption drifts around houses at Katmai, as shown in this photo taken two months after the June 1912 cataclysm.

Explosions and Explorers

We are awaiting death at any moment ... A mountain has burst near here, so that we are covered with ashes ... Here are darkness and hell, thunder and noise. I do not know whether it is day or night. ... Pray for us.
— *Ivan Orloff to his wife Tania, Kaflia Bay, 1912*

"No one suspected that such a catastrophe was in progress until the 6th of June..." Griggs wrote four years after the 1912 Novarupta blast, beginning his account of the disaster as something discovered and experienced from beyond Katmai shores, as if southwest Alaska were an unknown and unpeopled place. In truth, the Native residents of Katmai not only suspected but prepared, and their understanding of what was coming tells us volumes about this land's insistently dramatic personality.

For a week, increasingly severe earthquakes had shaken the coast, forecasting the upheaval to come. Though no major eruption had occurred in recent history, the Katmai volcanic cluster had experienced seven major explosions and many lesser ones in the last 10,000 years. Oral history had preserved an appreciation for the forces to be unleashed.

Many villagers were already on the Bristol Bay coast to the west, engaged in summer employment. Others fled in that direction in the days before the eruption or made perilous last-minute journeys, as in the case of Savonoski villagers who made a two-day trip in bidarkas to Naknek through hot, falling debris.

A small number of residents who had remained on the land or in coastal villages close to the eruption site observed the ashfall up-close. Their reactions were a mix of excited curiosity, especially on the part of children, and sober preparation on the part of adults who busied themselves storing water, overturning boats, and preparing to wait out the disaster. Meanwhile, even in towns far away, there was no doubt that great violence had been unleashed. In

Novarupta
The source of the 1912 explosion was not completely understood until the 1950s, when the eruption was tracked to Novarupta, a newly formed caldera possibly connected by a system of subterranean "plumbing" to the collapsed caldera at Mount Katmai, which Griggs originally suspected as the main eruption site. The Novarupta eruption released 30 times the volume of magma as the 1980 eruption of Mount St. Helens. Even the 1991 eruption of Mount Pinatubo, the century's second largest eruption, was less than half the size of the Novarupta eruption, which ended with the extrusion of a lava dome that plugged the Novarupta vent.

Explosions and Explorers

Katalla, 410 miles away and southeast of Cordova, the explosion sounded like "dynamite in the near-by hills."

When villagers like George Kosbruk emerged days later from a shelter in his fishing camp at Kaflia, they saw porpoises, birds, and fish floating on ash and pumice several feet deep. Kosbruk, 18 years old, was one of nine young men chosen to paddle three kayaks across Shelikof Strait to Kodiak for help. A borrowed cannery tug, the *Redondo*, returned to round up villagers. Rescuers noted the devastation, including streams so choked with ash that the water, turned into a cement-like slurry, had stopped running. In the wake of the disaster, many Native villages could not be reoccupied.

In more populous Kodiak, where the refugees would soon find shelter, the evening of June 6 turned dark with ash, but residents weren't disturbed at first, even while they were cut off from most information about the nature of the disaster. A radio station on Woody Island had been struck down by lightning during the eruption. The wireless apparatus onboard the docked United States Revenue Cutter *Manning* was useless because of the electrical condition of the air.

The following day, a second ashfall increased concern, and now both plans for evacuation and the spread of panic began. Church bells rang against the pitch-black afternoon sky. The saloons, which had been crowded, now closed, the proprietors agreeing that it was "a time for every man to keep his mind clear." When dawn failed to appear on the morning of June 8, nearly 500 residents were given refuge on the *Manning*, in a space meant for only a fifth as many people, creating conditions so packed that survivors later believed the overcrowding was worse than the eruption itself. But no one could have known that at the time, when it seemed that Kodiak was merely counting the hours until their Pompeii-like doom. No direct fatalities were recorded as a result of the eruption.

The Aftermath

At Kaflia on the Katmai coast, the Coast Guard reported ash to a depth of three feet. Closer to the volcano itself, up to 700 feet of ash was deposited over an area of 40 square miles. But that devastated landscape, well away from occupied Native village sites, would remain undescribed for several years.

The National Geographic Society sent seven expeditions to Kodiak and the Katmai coast between 1912 and 1919, four of them led by botanist Robert Griggs. In 1916, Griggs and his team of five men landed on the coast and traveled up the Katmai River valley, climbed Katmai Pass, and beheld for the first time what they named "The Valley of Ten Thousand Smokes," still actively steaming even four years after the volcanic blast. Vapor columns rose in pillars or merged into entire walls, blocking views about five miles down the valley.

Two members of a 1919 National Geographic expedition cook over a steaming Katmai fumarole. Griggs and his men marveled that they could cook bacon and flapjacks without any taint from the noxious gases entering their food. But one had to pick a "mild-mannered" fumarole; the stronger ones blasted with enough pressure to toss a frying pan into the air.

Fire and Ice

In Alaska, fire and ice are not incompatible. Some glaciers on Katmai's western side look like mud slides, covered with a layer of ash that, far from melting the ice, actually insulates it. Those glaciers predate the 1912 eruption, but a new glacier formed on a bench on the inside wall of the caldera of Mount Katmai after volcanic activity ceased that same year. It is one of the few glaciers in the world having a known date of origin.

Explosions and Explorers

A year later, Griggs returned, still "overawed" by the scenery, unable to contain his astonishment at what he considered one of the most amazing spectacles ever seen in nature. He bestowed the area with new names meant to convey the area's dramatic transformation. A valley that was nearly dry, its river vaporized when hit by the hot ash flow of 1912, would become home to a warm, muddy torrent named the River Lethe after the river that flows through the center of hell in Greek mythology. A mountain was named Cerberus after the three-headed dog that guards the gates of hell.

Struggling to convey the magnitude to readers unfamiliar with Alaska's vast landscapes, Griggs noted how damaging the eruption would have been in any place less remote. If centered upon New York City, for example, there would have been no survivors, and an area twice as big as Manhattan "would open in great yawning chasms, and fiery fountains of molten lava would issue from every crack." The sounds of explosions would have been audible in Chicago, and the fumes would have swept "all over the states east of the Rocky Mountains." Thanks solely to the remote location and lack of direct fatalities, Griggs and his readers could find awe in the demonstration of Earth's volatility without a morbid specter haunting their curious delight.

And what a delight it was, once the explorers got past their reasonable terrors of crossing a baking, gas-filled country. Pumice boulders were so light that they could be piled on top of a recumbent man, demonstrating the peculiar nature of air-filled rock. Around the camps, men experimented with cooking over fumaroles, destroying pots in the process or losing their sizzling bacon slices into the air as steam pressure built up against the bottoms of fry pans.

Though the valley had been decimated and buried in ash, and incomprehensible numbers of animals died in the disaster, wildlife was still in observance. According to Griggs, bears

Early explorers delighted in showing off the oddities of a volcanic landscape, including pumice boulders light enough for a man to lift. Today's park visitors can find plenty of pumice along Katmai lakeshores and in the Valley of Ten Thousand Smokes. Pumice's floating property is due to the thousands of tiny holes formed by the expansion of gases trapped in frothy lava as it cooled and hardened.

actually seemed attracted to the smoking valley, even without any food sources around. Basing his comments on the number of tracks seen at the edges of the largest vents, he said it was "not at all improbable" that they were like dogs seeking the warmest place to sleep near a fireplace. On successive trips, despite lessening activity,

Explosions and Explorers

Griggs and his explorers could still melt bars of lead by exposing them to the hot gases of open fumaroles of 645 degrees Celsius, or nearly 1200 degrees Fahrenheit.

Adventure-filled accounts by Griggs in *National Geographic* magazine entertained eager readers—a welcome distraction in an increasingly politically unstable world. In April 1917, the United States declared war on Germany. The following January, President Wilson delivered his famous 14 Points speech, outlining a plan for a peace treaty in Europe. In spite of this serious climate—or perhaps because of it—time was found in September 1918 for the President to set aside The Valley of Ten Thousand Smokes as a national monument called Katmai, a unit of the National Park Service, created only two years earlier.

Expanded by several presidential proclamations and administered by the distant Denali (then called Mount McKinley) National Park office, the monument was not visited on foot by a single park service employee between 1918 and 1940, by which time the thousands of smoking fumaroles had cooled to about ten. Griggs had thought the fumaroles were connected to a magma chamber with a long-lived heat source. Later scientists determined that most of the fumaroles were "rootless" and short-lived, with exceptions at the margins of Baked, Broken, and Falling mountains, and other locations near the Novarupta vent.

Today's Valley of Ten Thousand Smokes

The Ukak River valley was burned and buried by avalanche-like pyroclastic flows. Farther away, airborne ash suffocated vegetation and left a ghostly country. But almost immediately, faint signs of succession became visible. The study of this natural revegetation process in more distant Kodiak as well as devastated Katmai was one of the first goals of the National Geographic Society expeditions.

Despite the recovery of many parts of the larger park, well away from the eruption site, today's Valley of Ten Thousand Smokes still looks nearly as stark and unworldly as it did a century ago. In the Ukak River valley, layers of ash and pumice hundreds of feet deep cooled and were eroded by streams, leaving light-colored, sculpted river bluffs that invite comparisons to the canyon country of the American southwest. Bands of color—yellow, white, red, and black—bear traces of minerals deposited by the hot fumarolic gases that gave the smoking valley its post-1912 name.

If this mostly unvegetated landscape strikes the visitor as oddly lunar, NASA thought so, too. The space agency used the valley for trainings in 1965 and 1966. In the first year, the party of ten astronauts, including famous moon-walker Edwin "Buzz" Aldrin, were accompanied by geologists for hikes in the valley, where the fine-ground red pumice soil was thought to resemble what space mission candidates would find on the lunar surface.

Today, daily summer tours travel the unpaved park road from Brooks Camp to the Griggs Visitor Center, a 23-mile trip from boreal forest to alpine tundra. From the center, visitors can hike

Scant wildflowers brighten a stormy day in the Valley of Ten Thousand Smokes.

Explosions and Explorers

down to the edge of the valley floor, exchanging the birdsong and lush plants of the visitor center environs for the austere, ashen landscape of the valley.

With the power of Novarupta's eruption still very much in evidence, visitors may wonder if another major explosion is possible. The answer: a definite yes, given the active tectonic and volcanic nature of this landscape. Today's park is home to 15 active volcanoes, including the dense Katmai group which includes mountains Griggs, Katmai, Trident, Mageik, and Martin, which form a half-circle around the southwest to northeast sides of the Valley of Ten Thousand Smokes. Trident Volcano released several lava flows and small ash clouds in the 1950s. There is no historical record of eruptions at Mageik, Martin, or Griggs, but their actively steaming, sulfur-rich fumaroles suggest that a near-future eruption is possible.

The Alaska Volcano Observatory continues to monitor at least 30 active Alaska volcanoes of the more than 50 that have been active since about 1760. Compared to a century ago, the public is far more prepared for volcanic eruptions—but more vulnerable, too, given our reliance on air travel in particular (Anchorage has the largest amount of air cargo of any U.S. airport). This volatile northern segment of the Pacific Ring of Fire provides opportunities to study and marvel at the intense dynamism of the Earth, showcasing powerful processes often hidden from us.

Trained as a botanist, expedition leader Robert Griggs was first and foremost interested in observing revegetation processes following the 1912 Novarupta eruption, and the first National Geographic Expedition to nearby Kodiak was undertaken with that goal in mind. Lured closer to the eruption site on expeditions to follow, Griggs and his men traveled across a once-lush valley burned and buried beyond recognition. In 1916, Griggs found mosses and algae beginning to colonize moist areas near the steaming fumaroles. Fast-growing carpets of nitrogen-fixing liverworts followed, to be replaced as the fumaroles cooled and vanished with fireweed and horsetails, the latter able to force their way through three feet of ash, according to Griggs. In following years, grass, sedges, dwarf willow, and other hardy plants took hold in marginal environments with sufficient moisture and shelter from wind. Still, much of today's Valley of Ten Thousand Smokes remains barren, the ash deposits consolidated into hard-packed tuff, a type of rock. In places, the austere valley is cut through by fast-flowing streams, creating a unique post-eruption landscape of dramatic gorges and desert-like hues.

Aniakchak and Father Hubbard
Myth and Misunderstanding

One of the least visited of the national park units due to its severe weather and remote location halfway down the roadless Alaska Peninsula, Aniakchak National Monument and Preserve is wild and rugged even by Alaska standards. Few tourists will ever know this place firsthand; even those who explored it in decades past had trouble seeing beyond romantic ideas about the place's inaccessible, dramatic nature.

At the heart of the monument is the Aniakchak caldera, an ice-free crater six miles across and over 2,000 feet deep. It was formed a hundred years after the last great Egyptian pyramids were built. Those human monuments would appear small compared to the 7,000-foot mountain that collapsed as Aniakchak caldera was formed, losing about a third of its height by the time the debris settled. In the last 10,000 years, the area has seen at least 40 eruptions, the most recent in 1931.

At the caldera's northeast corner is the aptly named Surprise Lake, the most unexpected feature of which is a population of salmon that spawns in these warm spring-fed waters—an example of salmon's ability to recolonize newly created or dramatically changed habitats. The lake drains through a narrow breach in the caldera called The Gates, a steep-walled cleft through which the river races and winds howl, threatening to shred the equipment of the very few rafters who float this area each summer. The challenging trip takes three to four days from Surprise Lake to Aniakchak Bay on the Pacific Ocean.

While human visitors are relatively few, wildlife is not, as suggested by the names of area waterways, including "Cub Creek" and "Plenty Bear Creek." The tundra-covered landscape is thick with stands of willow and alder, hard traveling for people but little obstacle to the brown bears frequently spotted by rafters.

Between 3,400 and 500 years ago, the Aniakchak caldera wall ruptured, forming the cleft now called "The Gates" and catastrophically flooding and scouring the valley below.

Aniakchak was proclaimed a national monument in 1978, and Aniakchak River was named a national wild river two years later. The region through which this caldera-born river flows first came to national attention five decades earlier, however, when a Jesuit priest with a flair for publicity mounted a small expedition to its remote heart. Father Bernard Hubbard, head of the geology department at Santa Clara University, California, earned the moniker "The Glacier Priest" for his frequent journeys to places where the power of fire and ice were apparent—from Alaska's Mendenhall Glacier near Juneau to Aniakchak, a still-steaming node in the Aleutian Range.

Hubbard was not the first non-Native to describe Aniakchak caldera; various Russians and Americans, including some geologists, had previously explored the area. But he secured his place in southwest Alaska history thanks to books, articles, and films that spared no melodrama. Hubbard, who traveled the Aniakchak area in the company of several Santa Clara football players, made three trips between 1930 and 1932. While en route for the second expedition in 1931, Aniakchak erupted. For two months, Hubbard and his collegiate explorers had a chance to witness a landscape transformed. He described the wasteland that contrasted with the place he'd seen a year earlier: "There was the new Aniakchak, but it was the abomination of desolation, it was the prelude of hell. Black walls, black floor, black water, deep black holes and black vents; it fairly agonized the eye to look at it." On a third trip in 1932, Hubbard and pilot Frank Dorbandt made the first successful Aniakchak plane landing, in a floatplane on Surprise Lake.

While Hubbard's photographs and descriptions of the caldera provide an interesting benchmark, in other ways the "Glacier Priest's" desire to romanticize his own explorations worked counter to our understanding of the larger Aniakchak region. His emphasis on the alien and seemingly uninhabitable nature of this place contrast with what even he knew at the time: that the Alaska Peninsula was in fact a cultural crossroads rich in history—despite the area's severe climate and volcanic nature. As Katherine Johnson Ringsmuth writes in *Beyond the Moon Crater Myth*, the real Alaska Peninsula is "a place where Alutiiq shamans, Russian *promyshlenniki*, Orthodox priests, American traders, oil boosters, Inupiaq herders, salmon packers, fox farmers, trophy hunters, infantry men, and park rangers interacted at various points in time."

Father Hubbard charmed audiences with his gas-mask-outfitted dogs and descriptions of scorched landscapes.

A Land Both Harsh and Plentiful

A Land Both Harsh and Plentiful

Whether passing through Alaska or to more southerly lands, or settling in for centuries, people have always adapted to extremes: light and dark, water and ice, thick forest and windswept tundra, a glut of fish, game, birds, and berries, or none.

— Ellen Bielawski, archeologist.

Four thousand years ago, the clear waters of this lake churned with caribou, gathering at the narrowest place to cross. The "superlake"—Naknek, Brooks, and other Katmai lakes combined—was larger then, with water levels up to 85 feet higher than they are today.

Picture the massing herd on the shore, heads raised above the blue water, nostrils flaring; the clatter as hooves find purchase on opposite lakeshore. In the commotion and crush of bodies, it's easy to overlook the approach of hunters. An *atlatl* lifts and a dart is released, instigating the hunt that the combined forces of geology and biology have made possible by funneling so much life into one accessible place. At night, campfires blaze and eventually the land swallows up those traces to be found by future archeologists: scraps of caribou bone and chips of hunting tools, left by more than one of the culturally distinct societies that frequent the area.

Most of the Alaska Peninsula's human history follows the pattern visible at this singular spot: first, the reshaping of land by the master forces of glacial ice, seismic shifts, rising or falling water levels. Changing coastlines and newly formed rivers are colonized by salmon or inhabited by land mammals drawn to their particular features. Where fish and game are plentiful, people find them; in various diverse sites across Katmai National Park, evidence for human habitation dates back over 9,000 years.

There is bounty, but always in the wake of that bounty, there are additional challenges. New people arrive, forcing a competition for resources. The climate warms or cools, changing

Amalik Bay

Long before the rise of cities and the dawn of many civilizations, Katmai was an important cultural and natural dividing line, across which ideas and technological innovations were exchanged. The oldest known national historic landmark site on the Alaska Peninsula, Amalik Bay tells this ancient and fragile story, illuminated by artifacts dating back more than 7,000 years.

Amalik Bay, a large embayment that includes numerous distinct settlements and archeological sites, is located between the Bristol Bay side of the Alaska Peninsula, which currently is linguistically Yup'ik Eskimo, and Kodiak Island, which is Sugpiaq/Alutiiq Eskimo. Situated at a geographic boundary, the multiples sites also straddle a chronological and cultural transition. The area's lowest, oldest levels date precisely to the cusp between the Paleoarctic (mostly interior) and later traditions. Important artifacts from Amalik Bay include ground-slate tools and Norton-style pottery, hallmarks in the development of coastal Eskimo economies. These artifacts tell stories about early coastal adaptations, seasonal lifestyle patterns, and harvesting of animals over time—stories told in better detail here than at more recent sites—in a place notable for its dense concentration of prehistoric occupation. Amalik Bay was declared a national historic landmark in 2005. The National Park Service has undertaken projects to protect sites from coastal erosion that may be exacerbated by changing climate.

lifestyles over decades and centuries. Or in a single day—even a single hour—everything can change at once. The skies darken, the smell of sulfur fills the air, and cinders blanket the land. As multiple layers of ash attest, volcanic explosions have reshaped the land and lifestyles of the Alaska Peninsula again and again. What is perhaps most amazing is that the pulse of life—the caribou crossings, the pooling salmon, the quiet footsteps of hunters and fishers—has always managed to bounce back from or coexist with the swings of climate and the suffocating disruptions of volcanic events.

A snapshot from one time—over 20,000 years ago—would show enormous glaciers filling all of Cook Inlet, massive lobes spilling over the Alaska Peninsula, stopping just short of Bristol Bay. Another snapshot taken 10,000 years later shows the land those retreating ice age glaciers left behind: a long ridge of 7,000-foot volcanoes dividing the eastern coast of steep lands and short rivers plunging into the Pacific, and to the west, a flatter, lake-pocked land of marshes, rivers, and some of the most productive salmon habitats in the world.

The diversity and vastness of these landscapes, divided by a shield of volcanic ridges, would predict a diversity of cultural settlements, connected by waterways, passes, and trails. The archeological evidence supports this prediction in multiple locations across the park, from the coast to the interior. Although only about five percent of the park has been surveyed, it has been enough to reveal one of Alaska's densest concentrations of archeological sites, suggesting both the importance of the Alaska Peninsula to our understanding of northern human habitation and the immensity of what remains undiscovered and unknown.

At Brooks Camp, falling lake levels at what was once a caribou crossing site created the slowly-lengthening Brooks River and its famous falls, an obstacle for salmon and a gathering place for the bears that prey on them. No wonder, then, that such an optimal fishing site would also become a place for permanent settlement.

The natural creation of the falls coincided with the establishment of houses about 3,500 years ago by people, probably from the north, belonging to the culture archeologists call the Arctic Small Tool Tradition. In addition to caribou bones, numerous salmon teeth are found in these ancient settlers' hearths and floors, a sign of the increasing significance of the fish resource. But occupation of this special fish and game hub has not been continuous.

Archeological Evidence

Artifacts found in the Brooks River area include ground blades and "arrowheads" for hunting, ground slate ulus for processing salmon, stone adzes, chipped stone scrapers, oil lamps, and sewing awls. Many other kinds of tools were probably used but do not preserve in the acidic soils here, including grass and bark baskets, and other organic artifacts.

Villagers from Naknek stand in front of salmon drying racks in this photograph taken during the 1919 National Geographic expedition.

A Land Both Harsh and Plentiful

For several centuries during the Arctic Small Tool Tradition Phase, the area was apparently not visited at all, suggesting a fluctuation in wildlife populations. Two major deposits of volcanic ash also coincide with this time.

When people returned to Brooks River, they were from a different cultural tradition: the Norton people. These hunters and fishers also made pottery and heated and cooked with lamps filled with sea mammal oil, showing that they had contact with the Pacific coast. A change in house style around AD 1100—including the addition of a 'cold trap' that kept houses warmer—is one clue that a new cultural tradition, the Thule, was now dominant.

The Brooks River Archeological District, with more than 900 house depressions, is one of the single most densely concentrated archeological sites in North America. As an archeological district it is one of only five places in Alaska listed as both a national historic landmark and on the National Register of Historic Places. At Brooks Camp, the National Park Service has excavated and reconstructed a 700- to 800-year-old semi-subterranean house that helps tell the story of this resource-rich area, occupied and reoccupied by people of different cultural phases over millennia.

Traveling forward in time to the historic period of European contact and colonization, we find Alaska Natives in both permanent villages and seasonal camps across the Peninsula. When the Russians expanded their fur-hunting enterprise along the Alaska Peninsula coast, they encountered Sugpiak/Alutiiq living along the Shelikof coast and at Savonoski, east of Naknek Lake, as well as Yupiq people at Naknek, Alagnak, and other areas north. Trade between Savonoski and the coast was common, and at least until the 1890s, men from Savonoski sometimes traveled to the coast to participate with men from coastal villages in the sea otter hunt.

Life at the turn of the century was still a blend of new and old: rifles and a cash economy mixed with the use of single- or multiple-seat kayaks and subsistence hunting for fish, belugas, seals, and other marine mammals. While survival-oriented food gathering has always been a top priority for Alaska Peninsula people, time was also invested in the creation of artistic and religious objects, from masked spirit figures, to puffin rattles, to beautiful beaded headdresses made of arctic ground squirrel fur and caribou hair.

At the time of the 1912 Novarupta eruption, Katmai village had a trading post. Russian Orthodox churches were established in Katmai village, Savonoski, Kukak, and Douglas.

Pebbles incised with stylized human figures, which might have been used as game pieces or ritual items, have been found in the Brooks River area. Similar motifs appear on pebbles, rock art and other objects found in Kodiak, suggesting a connection between prehistoric peoples of both areas.

People lived in semi-subterranean sod houses. But change, both natural and cultural, was just around the corner. Ashfall and the temporary devastation of food resources, including fish and small game, caused the abandonment of traditional villages. As village chief "American Pete" Kayagvak lamented, "Never can go back to Savonoski to live again. Everything ash."

Some villagers relocated to a new coastal site 165 miles down the Alaska Peninsula called Perryville, named after K.W. Perry, the captain of the Revenue Cutter *Manning* that had transported Katmai refugees from Kodiak, where they'd been sheltered following the eruption. In Perryville, government-built frame houses replaced the more environmentally adapted, cold-resistant, semi-subterranean native dwellings. As the effects of the volcano waned, other challenges—including the 1918 flu epidemic and the absorption of abandoned villages into the boundaries of the new Katmai National Monument—took its place. Just as people persevered over the Alaska Peninsula's long history prior to the Novarupta eruption, they have persisted since, witnesses to the remarkable regeneration of fish and wildlife that followed in the disaster's wake. Despite "American Pete's" lament, fish populations rebounded within one to two decades. A subsistence salmon harvest continued at Brooks River until the 1960s. Coastal Perryville remains a seat of Alutiiq culture today. People with historic ties to Katmai also live in South Naknek, Naknek, King Salmon, Kokhanok, Igiugig, Levelock, Egegik, Chignik, and beyond. Dedicated to retaining connections to a land both harsh and plentiful, Katmai descendants retain close ties to their cultural homelands.

Subsistence food gathering remains important to many rural Alaska Peninsula residents, including Sugpiaq/Alutiiq descendants of the Katmai area who maintain their traditional fall harvest of "redfish"—spawned-out sockeye salmon with a low fat content that dries easily for storage.

A House through the Ages

In 1953, while digging a cellar for storage near Brooks Camp, workers discovered pottery, stone, and bone artifacts. These finds inspired the excavation of a nearby trench that yielded additional artifacts, marking the beginning of Katmai archaeological exploration and the recording of the first prehistoric site at Brooks River. More trenches and many more years of off-and-on research followed. Don Dumond, whose doctoral dissertation focused on 4,000 years of human history at this site, describes some of the major Brooks River findings in the National Park Service booklet, "Story of a House."

Among archeologists' early findings was the fact that different sequences of artifacts, suggesting distinct cultures, are found in sites on either side of the Aleutian Range before AD 1000. Following this time, a single society appears to have spanned the Peninsula, from the Naknek River drainage to the west, to the Pacific Coast sites to the east, and extending to sites on the Kodiak archipelago.

In the late 1960s, Dumond and students excavated several houses, searching for clues to how people lived over centuries, what they ate, and how they coped with extreme weather, from winter temperatures that could plunge to 40 degrees below zero to summer floods that forced prehistoric residents out of their subterranean homes and into seasonally-preferable above-ground tents. Artifacts at the site include ground blades and projectile points for hunting, slate ulus for processing salmon, oil lamps, and sewing awls. Animal bones found reveal the use or presence of a great variety of animals, including caribou, fox, porcupine, beaver, hare, ground squirrel, and dog, as well as many species of birds and fish.

Researchers also focused on why these prehistoric village sites were abandoned. A thick layer of ash called "ash C" that blanketed Brooks Camp and the surrounding region suggests a particularly disruptive event around AD 1300—perhaps ashfall from an explosion of Aniakchak Volcano, 150 miles southwest.

People continued to live at the Brooks Village until about 200 years ago, when population patterns were disrupted across the Alaska Peninsula. The area continued to be used as a seasonal fishing camp until the 1960s. One of the reconstructed prehistoric houses, a short walk from the Brooks Camp visitor center, is now a public exhibit, helping visitors imagine a subarctic life that has changed dramatically in some ways—and in other ways, not at all.

Surges and Seasons

Cool, wet summers are the norm at Katmai, especially in the rainiest months of August and September, when averages range from lows in the 40s to highs in the 60s.

Surges and Seasons

In all things of nature there is something of the marvelous.
  — Aristotle

A major volcanic eruption ejects ash that can travel halfway around the world, mixing subterranean ingredients from one side of the globe into the soils and water of distant lands. But volcanoes aren't the only sources of such explosive, change-making power.

Every year, salmon make their own truly wondrous surge, sending a dynamic pulse of life flowing up the rivers and lakes of Katmai country. That flow delivers the materials of life from the great, plankton-rich oceans into otherwise nutrient-poor freshwater systems, and from those fresh capillaries into the hungry land itself, where animals and plants rely on salmon to complete their own life cycles. It's a food web and seasonal marvel worth exploring in greater detail, to understand the close links between Katmai National Park, its flora and fauna, and its seascapes flanking opposite sides of the narrow Alaska Peninsula.

Five species of Pacific salmon are found in Katmai National Park, but the most abundant is sockeye, or red salmon, the foundation of the economy, ecology, history, and culture of southwest Alaska. Beginning in winter, sockeye hatch in freshwater streams and lakes throughout Alaska, emerging from the gravels in spring. Unlike other salmon species, sockeye delay the downstream trip, spending a year or two in freshwater, feeding on aquatic insects and plankton, before finally migrating out to the ocean. While other salmon species are also dependent on freshwater, sockeye are the most reliant on large lakes during the fry stage, which explains why the Bristol Bay region, home to the state's largest lakes, has the biggest sockeye runs in the world. In saltwater, the sockeye will roam the oceans for two to three years, putting on weight until they've reached an average of five to seven pounds.

Ready to reproduce and heeding the call of their freshwater origins, they migrate back upriver as adults. On that final,

Birding in Katmai
With a diversity of unspoiled habitats, from lake-dotted tundra to oceanic fjords, it's no surprise that Katmai can boast an equally diverse bird list of at least 172 species, including five species of water-loving loons, as well as grebes, swans, geese, and ducks; forest- and tundra-loving raptors, including eagles, hawks, and falcons; shorebirds from plovers to phalaropes; gulls, murres, murrelets, puffins; and many more.

Surges and Seasons

freshwater trip of several days, the sockeye stop eating. Once they reach their spawning grounds, the sockeye salmon's colors change, from bright silver to red-bodied and green-headed. Their bodies, once large and firm, become increasingly weak and ragged, with humped backs and elongated jaws. The survivors of this trip release eggs or sperm in a series of gravel nests called redds before dying. The young of the next generation will emerge the following spring, repeating the cycle. But the connection between one cohort of salmon and its offspring is only one vibrant strand on a shining web.

The Bristol Bay sockeye salmon fishery is the world's largest. Sustained as an industrial fishery for over a century—and a foundation of Alaska Native culture for thousands of years—it results in the capture of millions of fish yearly (27 million in 2010). About four million sockeye return each summer to Bristol Bay with the Naknek River system as a destination, just one of nine major river systems. Of these, more than one million reach their spawning grounds. While all Bristol Bay sockeye belong to one species, scores of distinct populations are adapted to the specific ecological conditions of individual rivers, streams, and tributaries. Variations in yearly conditions challenge some populations while boosting others. With so much diversity, the species as a whole—likened by researchers to a well-balanced financial portfolio—continues to thrive.

Whether or not individual salmon succeed in spawning, all contribute ecologically on their return journeys. Preyed upon by bears, scavenged by eagles, left to decay in streams and lakes, the salmon and their eggs end up feeding over 40 vertebrate species, including juvenile salmon, trophy-sized rainbow trout, wolves, mink, gulls, and magpies, as well as invertebrates, like aquatic insects that, in turn, are essential food for fish species.

One of the most invisible parts of this web is the way in which salmon—decaying on a streambank or carried by an eagle hundreds of feet or more inland—feed the region's plants and trees.

Salmon Roe

Pound per pound, salmon eggs (or roe) have seven times as many calories as salmon flesh, making them a special treat for bears and a critical food for rainbow trout, which reach world-class size in Katmai as a result of the healthy salmon fisheries of southwest Alaska. To consider how many prized calories each female fish carries, consider that she deposits more than 2,500 eggs, distributed across three to five nests. A male salmon travels alongside the female, depositing his sperm (called milt). Fanning her tail, the female covers the nest with a layer of gravel and generally lives only a few more days—never to see the juvenile fry that will emerge from gravel in spring.

Sockeye are the only Pacific salmon species that spend one to two years rearing in lakes, where they feed on aquatic insects and plankton, before migrating to the ocean.

Surges and Seasons

Scientists liken decaying salmon to a sack of fertilizer, providing phosphorus, nitrogen and calcium to low-nutrient systems. Nitrogen, in particular, is a limiting factor for terrestrial plant growth in northern and temperate forests. One study found that trees and shrubs near spawning streams derive about 22 to 24 percent of their foliar nitrogen from spawning salmon. By increasing streamside plant growth, which in turn keeps stream temperatures shaded and cool, salmon contribute in yet other ways to a positive feedback loop that benefits salmon as well as the many other animals in the southwest Alaska food web.

The Nature of Katmai

Katmai is so famous for its brown bears, salmon, and rainbow trout that it's easy to overlook the other species that thrive in a wild, geographically diverse area larger than any national park in the Lower 48 states.

The deeply indented, 497-mile coastline, shaped into fjords by ancient glaciers, is rich in marine species, including seals and sea lions, orcas and gray whales. Scientists estimate that 6,000 sea otters call the Katmai coast home, and a variety of seabirds also patrol the coast.

Most of the park is above treeline (less than 1000 feet in Katmai) and exposed to extreme weather. The plants that survive best here hug the ground in mats, with small flowers and waxy leaves that resist the drying effects of the wind. At home in open tundra and scattered woodlands are moose, caribou, red fox, wolf, wolverine, lynx, porcupine, snowshoe hares, and red squirrels.

On the road from Brooks Camp to the Valley of Ten Thousand Smokes, visitors have a chance of seeing some of those species, as well as the spruce grouse that take dust baths along the road itself. The fringe of vegetated landscape at the Griggs Visitor Center is made musical with the early summer songs of many birds, including Wilson's warblers, orange-crowned warblers, white-crowned and golden-crowned sparrows, hermit thrushes, and chickadees. Especially in June, avian melodies contrast with the more sterile landscape beyond, in the heart of the volcanic valley.

Southwest Alaska represents the marginal edge of Alaska's forests, a low-diversity terrestrial landscape with very few species of trees. Kenai paper birch, balsam poplar, and white spruce along with Sitka spruce are the only plants to regularly grow to tree size. Various species of willows and alders can also grow tree-sized specimens, but more routinely create dense tangles of shrub thickets. South of King Salmon, the gateway town just west of the park, very few spruce grow. A newcomer on the Alaska Peninsula, having arrived about 6,000 years ago, white spruce slowly expand their range southwest—reaching toward the windblown Aleutian Islands—at about one kilometer per year. Sitka spruce thrive only in limited northeastern drainages and along a thin band on the park's eastern coast. At the highest elevations in the Aleutian Range, snowfields and bare rock dominate the parts of the park that receive the fewest travelers of all. ∎

Bears aren't the only large mammals in Katmai willing to wade for their dinner. Moose feed on willows, grasses, and water plants, and are surprisingly strong swimmers. Wolves may also endure a cold splash in order to scavenge for salmon and have been seen in close proximity to brown bears at prime fishing sites.

Katmai From the Air

Like many of Katmai's first tourists, most of today's visitors first see the park, as well as the western gateway community of King Salmon, from the air. And what a view it is. From above, the flatter portions of the park and surrounding landscapes look like a brightly-hued sponge, dotted with lakes and smaller kettle ponds that are home to swans, ducks, loons, grebes, arctic terns, and lodge-building beaver. These vast and valuable wetlands resist draining, due to ice and clay underlying the soil. Low-elevation depressions and shallow-based lakes are bracketed by glacier-deposited moraines that have softened over time into rolling hills covered by the thinnest veneer of tundra plants.

Sculpted, scraped, and smoothed by the multiple advances and retreats of ice sheets between 8,000 and 25,000 years ago, the land has not lost all memory of its frozen past: scattered areas are underlain with permafrost, which is ground that has a temperature continuously below the freezing point for more than two years.

At higher altitudes, the traces of a frozen past have survived well into the present. Six percent of Katmai National Park is covered with glaciers. Along the park's eastern coast, some of these "rivers of ice" descend almost to sea level; the largest are three to four miles wide and ten to twelve miles long. The highest peaks of the Aleutian Range, which form a spine running down the Alaska Peninsula, rise to altitudes of greater than 7,000 feet, including the park's highest peak, Mount Denison (7,606 feet). Only a few passes, including Katmai Pass, offer low-relief routes through these icy mountains into the flatter country to the west.

An Industrious and Alluring Frontier

An Industrious and Alluring Frontier

The fishing in these lakes and rivers make the region an angler's paradise. Their waters are alive with giant rainbow trout, with such voracious appetites that the angler never need cast more than once or twice before he has a strike that keeps him busy.
— Robert Griggs, 1922

While the appeal today of Katmai for many is its thousands of square miles of undeveloped wilderness, it would be misleading to imagine this country was not home to diverse attempts at development and resource extraction, beginning with the arrival of Russians to the upper Alaska Peninsula in the 1760s.

Shelikof Strait, off the park's eastern coast and named for Russian fur merchant Grigorii I. Shelikhov, is a reminder of that century-long era, during which rival Russian companies vied for domination of the fur trade, pressing Katmai Natives into their service from the coast to Savonoski. Sugpiaq/Alutiiq hunters from Katmai were forced to join larger contingents that traveled as far as Sitka, then capital of Russian Alaska, with a few continuing even further, to Fort Ross, California. Russian expeditions also explored other natural resources and mapped the coastline in detail. Katmai continued to be the primary fur trading post and population center even after the American purchase of Alaska from Russia in 1867.

Forcibly included in this emblematic Alaskan experience, Katmai is similarly connected by history to most other classically Alaskan experiments and ventures. Nearly everything that was tried in the rest of Alaska was tried here, too—from reindeer herding to gold prospecting. Then, too, there have been industries not common elsewhere, including clam digging and a small and ultimately unsuccessful attempt to mine pumice on Takli Island and Geographic Harbor.

At the turn of the century, Katmai Pass became a shortcut across the Alaska Peninsula for some goldseekers en route to Nome. A different gold strike in Kuskokwim—which turned out to be

Brooks River is world-famous for its rainbow trout population and also attracts anglers seeking grayling, silver and sockeye salmon, Dolly Varden, and arctic char. Shallow enough for wading from end to end and side to side, Brooks is perfect for flyfishermen—especially ones willing to get moving and make room for the brown bear intent on any particular sandbar. (The abundance of bears is one reason trout anglers will find the best river fishing before the salmon run begins, in late June, or after it has petered out, in late August.)

bogus—was the destination for another trekker, the author Rex Beach, who wrote a memoir in which he briefly described the crossing from Katmai village to the Naknek region.

The million-dollar salmon business now so essential to Bristol Bay has a long history dating back to the 1880s, when canneries and salteries were first established. By 1900, the Naknek River canneries employed 31 local Natives and many more non-Alaskans, including 271 Chinese.

The boom in Alaska salmon processing hit just in time to offset the decline of the sea otter hunt. It was paired, in Katmai, by a quirkier and ultimately shorter-lived industry: razor clam canning. Clam harvesting began on the west side of Cook Inlet and Shelikof Strait in 1919, outside the boundaries of the monument that had been established just one year earlier. When the monument expanded, it absorbed both clam canneries and prime clamming beaches. A fire destroyed one cannery; another was allowed to continue operating within monument boundaries to supply the U.S. Army.

Fating it to become just a curious footnote in Katmai's industrious past, many problems plagued commercial Katmai clamming: competition from more mechanized East Coast operations, variable harvests due to severe weather, and shellfish poisoning concerns, which completely shut down all clam harvesting in the 1960s. Each time, the minor clam industry rebounded, and by 1971, 200,000 pounds of razor clams were being harvested per year in the monument. But for the next 15 years or so, until the commercial harvest ended, it remained a small-scale, rustic enterprise—one that left a fair amount of litter in its wake. In the early 1970s, a long string of shacks, abandoned cars and debris characterized the coast until a clean-up operation helped restore order to the beach. Clam permits were restricted to limited individuals until the early 1980s. Today, no clam industry exists in a park now far better known for its more charismatic megafauna.

The Sportfishing and Tourism Era

In contrast with clamming, one industry—fly-in sportfishing and the associated lodge trade—did much more than just coexist with the monument and later national park. It helped bring Katmai to the world's attention in the first place.

The explorer Robert Griggs was lured by Novarupta's devastation and the scientific lessons it could impart. But even he noted that Katmai country beyond the ashen wastelands had much else to offer, including stunning scenery and spectacular fishing

Clamming

Razor clams are found on sandy tidal beaches from the Bering Sea to Southern California. Of the eight known major concentrations of clams on the Pacific Coast, four are in Alaska. Maximum size for the highly edible, soft-shelled clams is over 12 inches; most mature clams are about half that size.

The Naknek Packing Company dock, as seen by the members of the National Geographic expedition in 1918. The first salmon cannery opened on the Naknek River in 1890. Over 120 years later, the local economy is supported almost entirely by salmon commercial fishing and sportfishing.

An Industrious and Alluring Frontier

for trout so big and heavy that the fish broke all his men's tackle. (Griggs admitted that he and his men were novice anglers.) Griggs and many other travelers who visited just before or during the monument's early years predicted Katmai would one day be a tourist destination, with just as much to recommend it as any Yosemite.

Even after Katmai National Monument was expanded in 1931, it remained hard to reach except by small plane, making it off-limits to all but the most adventurous of tourists, yet still accessible enough to raise concerns about illegal hunting, trapping, and fishing. Despite increasing problems, enforcement was slow to arrive. Between 1918 and 1940, the National Park Service made only one single-day flyover patrol, in 1937. Without enough budget for staffing, the park had to rely on cooperative management with other agencies to patrol Katmai for many years.

Destined to slip into obscurity even as it rebounded ecologically from the 1912 eruption, Katmai made it back onto the visitor map during World War II and the years just following. During the war, which transformed Alaska's infrastructure and greatly increased the territory's population, Katmai became a popular fishing destination for off-duty military personnel from Naknek Air Force Base in nearby King Salmon. Anglers caught and kept uncounted numbers of rainbow trout.

Enter aviation pioneer Ray Petersen, who in 1947 had founded his Northern Consolidated Airlines using surplus DC-3 war aircraft. Petersen, who was committed to fishing conservation, proposed to establish five "Angler's Paradise" camps. The National Park Service granted him a five-year permit to build and operate those camps, making him the park's first concessioner, with John Walatka as the company's superintendent of camp operations.

Petersen provided both the places to stay and, via his airlines, the means to get there. But he went further, promoting southwest Alaska by arranging visits by outdoors' writers from national newspapers, as well as politicians and other public figures, including former presidential candidate Adlai Stevenson and *National Geographic* editor Gilbert Grosvenor, after whom one of the camps (originally named Coville) was later renamed in honor of the editor's 1954 visit. The first camps featured tent cabins and bunk beds—as well as some of the best fishing in the world. Early on, Petersen hoped to attract not only anglers, but also the everyday tourist "who goes places just to be amazed."

Broadening the appeal of the park to non-anglers, the 23-mile road to the Valley of Ten Thousand Smokes was built in 1962. That year, 350 people visited. By 1970, the number had swelled thirty-fold, to 10,000 visitors. Today, Katmai receives over 25,000 visitors annually.

Only catch-and-release fishing is allowed for rainbow trout; consult park information for more specific dates and restrictions.

An Industrious and Alluring Frontier

The equation of many more people plus a recovering population of increasingly bold brown bears was bound to add up to problems. By the 1960s, it did. A non-fatal attack and a pattern of bears preying on garbage and food prompted the development of the park's first bear management plan. Over the years, precautions were implemented, from the closing of open dumps to a new policy of incinerating garbage to the refinement of visitor safety talks.

In 1980, President Jimmy Carter signed into law the Alaska National Interest Lands Conservation Act, which doubled the size of our country's National Park System, expanding many existing parks and creating new ones. Among these was Katmai, which was both expanded and also designated as a national park and preserve. Of the total Katmai parkland, over 3.4 million acres are now designated wilderness.

Since the 1980s, management of bear-human interactions and expanded research activities continue to shape the future personality of the park. While a third of visitation is focused at Brooks Camp, the rest is spread across other park locations, including popular angling rivers and the coast, where an increasing number of bear-watchers head in spring and summer.

Even as more people discover Katmai, it remains relatively little-visited when compared to other national parks. Nearly twice the size of Yellowstone, it receives only one percent as many visitors. What those visitors experience stands out by any measure: from the drama of watching brown bears feed and interact in the wild, to the experience of fishing pristine salmon and trout streams, to the wonder of seeing a valley transformed by the twentieth century's largest volcanic explosion. In a state known for its superlatives, Katmai has proven an alluring—and enduring—destination.

Changing Attitudes About Bears

In the park's first years, as the world struggled to recover from World War, the scarcity of beef caused many Alaskans to call for a repeal of all protection for brown bears. Katmai promoters warned that, for strategy's sake, given the era's anti-bear climate, the word bear should never be mentioned in connection with the establishment of a national monument.

Fure's Cabin

Roy Fure, originally from Lithuania, came to Naknek Lake's Bay of Islands in 1926. He trapped game, raised a family, and continued to use the cabin he built, plus another on American Creek, until his death in 1962. Fure's struggle to maintain the Naknek Lake cabin survived the extension of Katmai National Monument boundaries in 1931, the failure of an application for a homesteading permit, and a 1941 arrest for trapping within the monument. Fure was finally granted citizenship and willed the cabin to his daughter, who died in 1980, leaving no heirs. The renovated cabin, used today by Savonoski Loop paddlers was added to the National Register of Historic Places in 1985—a reminder of even lesser-known histories of people who made their living on Katmai land, including not only trappers and hunters, but fish and clam processors, fox farmers, and other hardy pioneers of many nationalities, for whom Katmai was not only a place of renewal and escape, but an industrious last frontier.

Alagnak River

The Yup'ik word "Alagnak" means to err or make mistakes—a fitting name for the many braided sections of this clear-watered, remote waterway, where the vast number of glinting, shallow channels complicate way finding. Nonetheless, people have found their way here for nearly as long as humans have been in Alaska. The river's human history dates back 9,000 years, according to archeological evidence of riverside camps. Even in modern times, the river retains its timeless qualities, with no road access or impediments and clean, unpolluted water that are home to rainbow trout, salmon, arctic char, arctic grayling, and northern pike.

Originating in the Aleutian Range, within Katmai National Park boundaries, and emptying, 85 miles later, into the Kvichak River near Bristol Bay, the Alagnak River is the most popular fly-in fishing river in southwest Alaska. Its upper portion, 67 miles long, is one of 26 congressionally designated national wild rivers. The river's journey begins gently as it winds through tundra, then quickens as the Alagnak is funneled between steep, spruce-forested cliffs, producing Class I to III rapids before slowing again to long, braided stretches dotted with numerous islands. The river travels through country that is home to brown bears, moose, caribou, and bald eagles.

Braided across a wild landscape, the river is also braided with the region's human history. Subsistence use of the Alagnak attracted not only Yup'ik people from the Kvichak River, but also people from as far away as the Yukon and Kuskokwim drainages. In 1900, two salmon canneries were built near the junction of the Alagnak and Kvichak rivers. The names of settler families from this era reveal a blend of cultural influences, bringing together Native Alaskans with European Americans of various backgrounds, including French-Canadian, Irish, and Scandinavian.

A significant event for the Bristol Bay region was the first airplane landing in June 1927, when pilot Russel Merrill of Anchorage Air Transport flew Libby, McNeil & Libby Company and Nakat Packing Company executives to the Lockanok cannery at the mouth of the Alagnak River. Connecting the region via air travel opened a new era for the otherwise isolated communities. Sportfishing may have started as early as the late 1930s and gained a world-class reputation beginning in the post-World War II era.

Branch River village, three miles up from the river junction, was the last historic settlement on the river and was abandoned by the 1960s. Most families had moved to other nearby communities. But family and cultural ties to this beautiful and wild river remain strong. Modern Yup'ik, Sugpiaq/Alutiiq, and Denaina people continue to make use of the Alagnak region for subsistence fishing, hunting, berry picking, and firewood gathering.

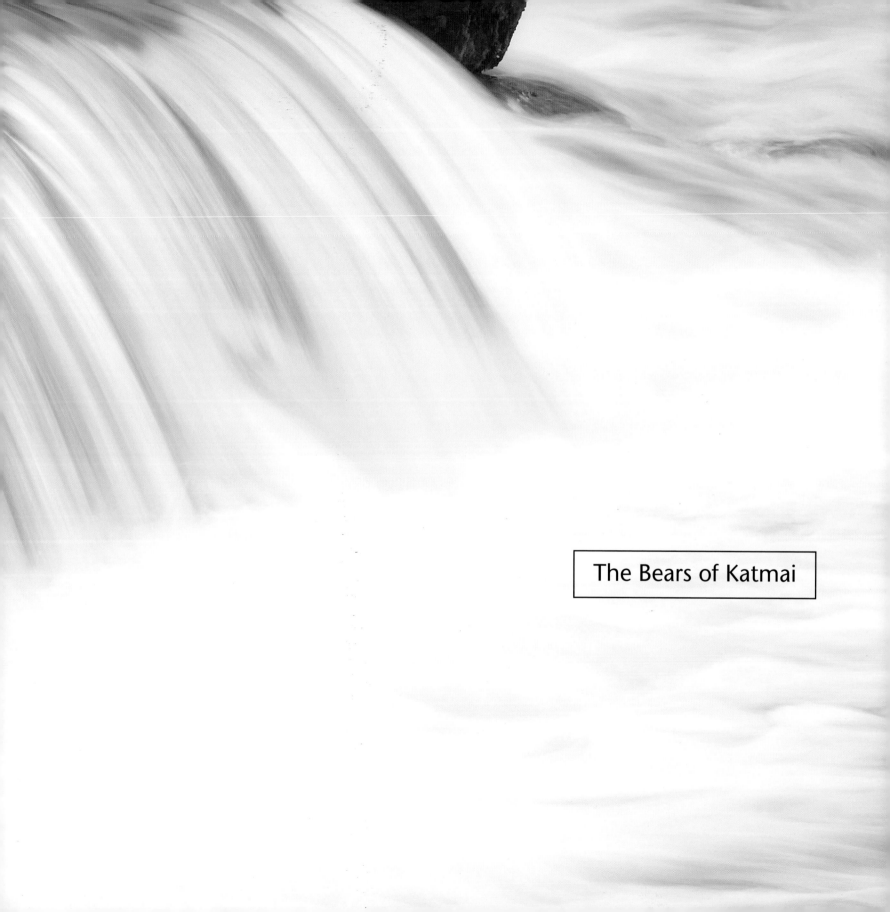

The Bears of Katmai

The Bears of Katmai

If we can learn to live with bears, especially the grizzly, and if we can learn to accommodate the needs of bears in their natural environment, then maybe we can also find ways to use the finite resources of our continent and still maintain some of the diversity and natural beauty that were here when Columbus arrived.

— *Stephen Herrero*

Coastal brown bears and interior grizzlies are both members of the Ursus arctos *species. The coastal brown bears' much larger size is attributable, in part, to the abundance of salmon in coastal and near-coastal areas. Black bears in Katmai are rare.*

Power and patience. The two coexist at the Brooks River Falls, a modest cascade of foaming waters bounded by thick forest and occupied, on this summer day, by no fewer than a dozen large brown bears. Famous photographs often show the bears standing with jaws open, making improbable mouth-first catches of leaping sockeye salmon. But the bears have more than one pose. Sometimes, they sit along the margins of the falls, fishing with their claws, stripping the bright-orange fish flesh with astonishing ease.

Or, the big bears sit with mouths closed, heads slung low, expressions wary as they guard a favored spot. On this day, one bear wallows persistently in the bitterly cold "jacuzzi," a bubbling hole at the waterfall's base. Suddenly, a blurry flash of motion disrupts the scene. A large male bear challenges another, bodies rise up, teeth glint, the growls and roars of ursine indignation sound above the rush of water. The loser retreats quickly, backing away from the falls to fish slightly downriver at a less contested spot.

The surprise of this scene is the number of bears present and the fact that they tolerate each other—most of the time. Bears are not herd or pack animals. A brown bear's home range can easily extend over a hundred square kilometers. Bear density in Katmai is among the highest in the world, with more than 2,000 brown bears estimated to live in the park. Denali National Park, by comparison, is home to 200 to 300 grizzly bears. While bear density at Katmai is about two-thirds that found in a comparable rich coastal

The Bears of Katmai

environment like Kodiak Island, it is ten to a hundred times the density of some Interior Alaska grizzly bear populations.

The difference between places that have lots of bears, and places that have relatively few, boils down to one essential factor: salmon. In Katmai, abundant salmon means abundant brown bear populations, and especially so in those special places where quirks of geography—like a waterfall that temporarily halts salmon, forcing them to pool before making the leap upstream—bring predator and prey together in exceptionally close quarters. Tolerance is the result, but even tolerance has limits, and smaller or weaker bears experience the consequences when they push their luck.

A Cub's Life

Katmai has prime bear habitat and one of the highest known densities of brown bears in the world, but that doesn't necessarily make it an easy place to be a bear cub. Only one-third of cubs will survive to adulthood, and predation by adult male bears is not uncommon. Cubs, most often born in litters of two or three, represent a great investment of maternal energy. A mother bear doesn't have her first litter until the age of eight or nine years old, and she spaces her litters three to four years apart.

Since large males have no objection to preying on cubs, mother bears will generally avoid the threat, preferring less-optimal fishing areas away from the noisiest and most dangerous territory contests. Sub-adult males, ages two and a half to five and a half years old, will work their way toward better spots as they gain in heft and confidence. Old survivors often have visible scars to prove the cost of defending their turf.

The result of these ages and interactions is a "fluid hierarchy" of boars, sows, and cubs balancing the risk of injury with the benefit of more calories from a maximum daily fish catch that for most bears runs to 10 to 20 salmon per day. With fish in such abundance, bears can choose to be a little picky, stripping some salmon for their favorite parts, including the fatty skin, eggs, and brains. Nourished with such high-fat fare, male brown bears can grow to 1500 pounds, or more than three times the size of interior grizzlies.

As variable as a bear's own place in the hierarchy is its own approach to catching fish. Some are generalists, able to succeed at more than one spot. Others specialize in using a favorite location or fishing in a preferred way, or by pioneering a less-common strategy, like diving for fish with heads fully submerged. Scientists and park staff make an effort not to anthropomorphize bears, but the aggregation of so many large animals in one place, and the variability in appearance and behavior, lends itself to nicknaming. The use of number codes to identify bears can't prevent people from learning to recognize and fondly remember "Headbob," "Backbite," "Snaggletooth," "Genghis," or "Battleship Flo."

In the early 1950s, when Willie Nancarrow became the park's first ranger, few bears were seen in the Brooks River area. A visiting reporter was escorted to Savonoski River to look for bears, suggesting that none were to be seen close-by, and camp records throughout the decade suggest that bear sightings were a rare occurrence. Five decades later, Brooks River is one of the top

bear-watching spots in Alaska, and about 70 bears are identified as regularly using the Brooks River—a number that has more than doubled since the late 1980s. Why the change? The population may be rebounding from earlier overhunting, harassment, or natural changes, like the 1912 eruption. Better fisheries management and improved salmon runs may be supplying enough food to support more bears. Population changes and the dynamics of bears feeding in close proximity with respectful humans will continue to inspire interest and research in years to come.

The Seasonal Round and Other Places to Watch Bears
Understanding bears' favored food sources is the key to bear-watching, with month-to-month variation according to salmon runs and other geographically-specific considerations.

In July, Brooks River Falls, the most visited area of the park, is the prime place to see bears vying for favored fishing spots. A second, quieter opportunity for watching bears occurs in September. During that month as the sockeye runs die down, many Brooks Camp bears are seen lower in the river and along the Naknek Lake shore "snorkeling" for the remains of spawned-out fish. While brown bears are considered terrestrial mammals—as compared to polar bears, which are actually classified as marine mammals—they seem astonishingly comfortable wading, swimming, and feeding in water.

Bears are numerous at other locations, too. In June, wildlife-watching trips more commonly focus on the park's eastern coast at places like Swikshak Lagoon and Hallo Bay, where bears are seen foraging sedges and grasses above the tide line, and digging clams below it. Wide, exposed beaches are the best places for bears to put their large, clam-digging claws to use, but the bruins will also wade into deeper water, heads under the surface as they unearth high-protein shellfish.

Katmai Fossils
Bears aren't the only giants to have walked this land. Dinosaur footprints found in the Chignik Formation, which covers a large part of Aniakchak National Monument and Preserve, are contributing to a slowly-emerging portrait of Alaska's high-latitude dinosaurs. Most northern Alaska discoveries have been from the Cretaceous period, and so too with the 2000 discovery of tracks from a hadrosaur, or duck-billed dinosaur—the first dinosaur fossils found in any Alaska national park. Adding to the discovery was the preservation of fossilized leaf litter marked with the feeding trails of herbivorous insects, providing evidence of a large standing forest. Future discoveries in both Aniakchak Monument and Katmai National Park promise to yield many more clues to Alaska's paleontological past.

During an hour of foraging, bears move about a half-ton of wet sand and consume an average 117 shellfish, according to Katherine Johnson Ringsmuth's *Buried Dreams*, a book dedicated to the delightfully esoteric subject of Katmai clamming history. Anyone who has dug hard for razors will understand that this ursine ability, though perhaps less famous and photogenic than mouth-first salmon fishing, is equally impressive, and perhaps just as nutritionally essential, given that in a single tidal cycle, bears typically will clam-dig for about three hours and net about 36 pounds of food, according to biologists.

In August, coastal bears congregate at rivers and streams to catch salmon. Geographic Harbor and Moraine Creek/Funnel Creek are popular bear-watching spots during that month.

As berries ripen, bears will make use of those calories as well, leaving a trail of berry-filled scat through the forest and across the tundra-covered hills. As fall turns to winter, protected park slopes provide important denning habitat. During the long and hungry season of winter, cubs are born—so vulnerable and at one pound, so small, it's hard to imagine they'll take their place fighting for the best fishing spots in just a few years' time. ■

Despite their diminutive size, berries provide bears with a source of nutrition in late summer and early fall.

Bear-Watching
Rules and Tips

A small floatplane lands, pulling up to the beach just as the grasses part and a brown bear emerges, taking its morning constitutional along a stony stretch already occupied with ursine neighbors. In one direction, a bear swims just offshore, paddling after sluggish late-season salmon. In the other direction, a warm gravel bar has become the loafing zone for a relaxed mother bear and two cubs. Before any of the bears can get any closer, visitors are guided into the park visitor center for a required bear safety training. Key instruction: Let bears have the right of way and approach no closer than 50 yards, a distance that feels close enough when a relaxed bear is ambling your way, demonstrating complete confidence that the beach, forest, and tundra all belong to him.

Next stop: the bear viewing platforms—but not without possible delay. Several park rangers with radios ask visitors to pause while a bear takes its time swimming around and under a bridge near the Lower River platform, close to the pedestrian trail. Though the most famous and photographed bear-viewing site is Brooks River Falls, a one-mile walk from the visitor center, in truth every place in Brooks Camp is a bear-viewing site, and even the shortest of walks between lodge buildings may require an unexpected detour to avoid crowding a bear passing through.

The National Park Service and visitors cooperate in keeping bears and people safe. They also cooperate in keeping bears wild by not allowing bears to steal food, including salmon at the end of an angler's line, given that "easy meals" can lead to food conditioning. Picnicking is permitted in designated areas of Brooks Camp. Within one and a half miles of the falls, camping is allowed only in an established campground, with rules for food and gear storage.

Despite these precautions and a general effort to give bears their space, visitors may have the experience of observing the behavior of bears under stress caused by human or bear proximity. While a relaxed bear is often too busy eating one salmon after another to make any sound other than chewing, stressed bears will vocalize. Tension may begin with yawning, especially if one bear feels another bear—or a person—is getting a little too close. Popping sounds, a series of huffs, or a single sharper exhalation called a "woof" are even clearer signals of discomfort. Mother bears may also use these sounds to warn or call their cubs. Growls are at the farthest end of the aggression continuum and may accompany a quick tussle between bears over fishing territory, with one bear forced to retreat.

If a bear rears up on hind legs, it may be sniffing the air, or just trying to get a better view. Back away slowly. If a bear continues to approach, talk in a calm voice and wave your arms. Never run from a bear. Most charges are bluff charges, and bears sometimes approach within feet of a person before veering off. If a bear does make contact—an exceptionally rare occurrence in Katmai National Park—fall on the ground, protecting your face and neck. If an attack is predatory or prolonged rather than defensive in nature, fight back vigorously.